INVESTIGATE!

MYSTERIES OF SPACE

DO OTHER EARTHS EXIST?

JENNIFER LOMBARDO

Enslow PUBLISHING

Please visit our website, www.enslow.com.
For a free color catalog of all our high-quality books, call toll free 1-800-398-2504 or fax 1-877-980-4454.

Cataloging-in-Publication Data
Names: Lombardo, Jennifer.
Title: Do Other Earth's Exist? / Jennifer Lombardo.
Description: New York : Enslow Publishing, 2023. | Series: Mysteries of Space | Includes glossary and index.
Identifiers: ISBN 9781978532830 (pbk.) | ISBN 9781978532854 (library bound) | ISBN 9781978532861 (ebook)
Subjects: LCSH: Exoplanet–Juvenile literature. |Life on other planets–Juvenile literature. |Astronomical telescopes–Juvenile literature.
Classification: LCC QB820 L66 2023 | DDC 523.2–dc23

Portions of this work were originally authored by Michael Portman and published as *Are There Other Earths?*. All new material this edition authored by Jennifer Lombardo.

Published in 2023 by
Enslow Publishing
2544 Clinton Street
Buffalo, NY 14224

Designer: Tanya Dellaccio
Editor: Jennifer Lombardo

Photo credits: Cover (logo) AAVAA/Shutterstock.com; series background Molibdenis/Shutterstock.com; Cover (Earth) robert_s/Shutterstock.com; p. 5 (bottom) ValentinaKru/Shutterstock.com; p. 5 (top) AstroStar/Shutterstock.com; p. 7 (bottom) ESA/Hubble & NASA; p. 7 (top) https://upload.wikimedia.org/wikipedia/commons/6/67/Artist's_impression_of_exoplanet_orbiting_two_stars.jpg; p. 9 NASA, ESA and the HST Frontier Fields team (STScI), Acknowledgement: Judy Schmidt; p. 11 (top) OSweetNature/Shutterstock.com; p. 11 (bottom) Viacheslav Lopatin/Shutterstock.com; p. 13 alionaprof/Shutterstock.com; p. 15 ESO/Bohn et al.; p. 17 (bottom) STScI/NASA; p. 17 (top) Russell Croman/NASA; p. 19 lovemushroom/Shutterstock.com; p. 21 NASA/CXC/Univ of Toronto/M.Durant et al; Optical: DSS/Davide De Martin; p. 23 Dotted Yeti/Shutterstock.com; p. 25 (bottom) NASA; p. 25 (top) Rishad Allaberdiev/Shutterstock.com; p. 26 Dima Zel/Shutterstock.com; p. 27 BEST-BACKGROUNDS/Shutterstock.com; p. 29 Diego Barucco/Shutterstock.com.

Printed in the United States of America

CPSIA compliance information: Batch #CWENS23: For further information contact Enslow Publishing at 1-800-398-2504.

CONTENTS

SO MUCH TO DISCOVER 4
LUCK AND SKILL 6
USING WHAT WE KNOW 8
WHAT TO LOOK FOR 10
FINDING THE RIGHT SPOT 12
HARD TO FIND 14
WHAT CAN DIRECT IMAGING TELL US? 16
LOOKING INDIRECTLY 18
A HAPPY ACCIDENT 20
OTHER JUPITERS 22
SPACE TELESCOPES 24
STILL LOOKING 28
GLOSSARY 30
FOR MORE INFORMATION 31
INDEX 32

Words in the glossary appear in **bold** type the first time they are used in the text.

SO MUCH TO DISCOVER

Imagine you're looking at the world around you through a cardboard tube. You can see a little bit, but you know there's much more out there. That's what looking at the night sky from Earth is like.

The invention of the telescope—a tool that makes faraway things look closer—helped us learn more about the **universe**, but it didn't give us all the answers. It wasn't until 1992 that scientists learned there are other planets outside our solar system! Today, we know there are trillions of planets out there. We call these exoplanets.

A TELESCOPE CAN GIVE YOU A CLEARER PICTURE OF THE NIGHT SKY. SCIENTISTS HAVE VERY POWERFUL TELESCOPES TO LOOK FOR EXOPLANETS.

EXPLORE MORE!

UNTIL THE 15TH CENTURY, SCIENTISTS SAID EARTH WAS THE CENTER OF THE UNIVERSE. THIS WAS CALLED THE GEOCENTRIC MODEL. ("GEO" MEANS "EARTH.") TODAY, WE KNOW OUR PLANET ISN'T EVEN THE CENTER OF OUR SOLAR SYSTEM! IT'S JUST ONE OF MANY OUT THERE.

LUCK AND SKILL

The universe is huge, and it's expanding, or getting bigger, all the time. We can only see a tiny part of it, and that part is filled with stars and planets.

Scientists **estimate** that there are more than 1 septillion (a 1 followed by 24 zeroes) planets in the **observable** universe alone—and those are just the ones that circle large stars! This makes finding another planet like Earth very difficult. Scientists will have to be looking at the right part of the sky at exactly the right time to even know it exists.

THIS PICTURE SHOWS WHAT ONE ARTIST THINKS AN EXOPLANET WITH TWO SUNS MIGHT LOOK LIKE.

EXPLORE MORE!

EVEN THOUGH IT HAS "YEAR" IN IT, A LIGHT-YEAR ISN'T A MEASURE OF TIME. IT'S THE DISTANCE LIGHT TRAVELS IN ONE YEAR. SCIENTISTS THINK IT WOULD TAKE LIGHT ABOUT 93 BILLION YEARS TO CROSS THE OBSERVABLE UNIVERSE!

USING WHAT WE KNOW

A galaxy is a group of stars, planets, and other objects that are held together by **gravity.** The universe contains trillions of galaxies that each have billions of stars. Scientists can estimate this number by multiplying the stars we can see in our own galaxy, the Milky Way.

When scientists look for other Earths, they start with what they know about our planet. This helps them narrow down their search to exoplanets that are rocky, in **orbit** around a star, and in the right spot to have liquid water.

EXPLORE MORE!

SCIENTISTS BELIEVE THAT AT THE CENTER OF EVERY GALAXY IS A BLACK HOLE. THIS IS A LARGE AMOUNT OF MATTER SQUEEZED INTO A VERY SMALL SPACE. THIS IS WHAT CREATES THE GRAVITY THAT HOLDS THE GALAXY TOGETHER.

EACH OF THE DOTS IN THIS PHOTO IS A GALAXY.

WHAT TO LOOK FOR

Scientists look for rocky planets because many living things on Earth need solid ground. This rules out planets that are like Jupiter, Saturn, Uranus, and Neptune. These gas giants don't have a solid surface but may be solid at their core, or center. They're mostly made up of a swirling mix of hydrogen and helium gas.

Scientists also look for planets that orbit a star because we know we need our sun to give us light and heat, so it's likely other kinds of life need that too.

EXPLORE MORE!

IT'S POSSIBLE THAT THERE ARE LIVING THINGS THAT HAVE **EVOLVED** TO LIVE ON GAS GIANTS OR PLANETS WHERE THE WATER HAS ALL BOILED AWAY. SOME SCIENTISTS ARE LOOKING FOR LIFE ON EXOPLANETS THAT ARE NOTHING LIKE EARTH.

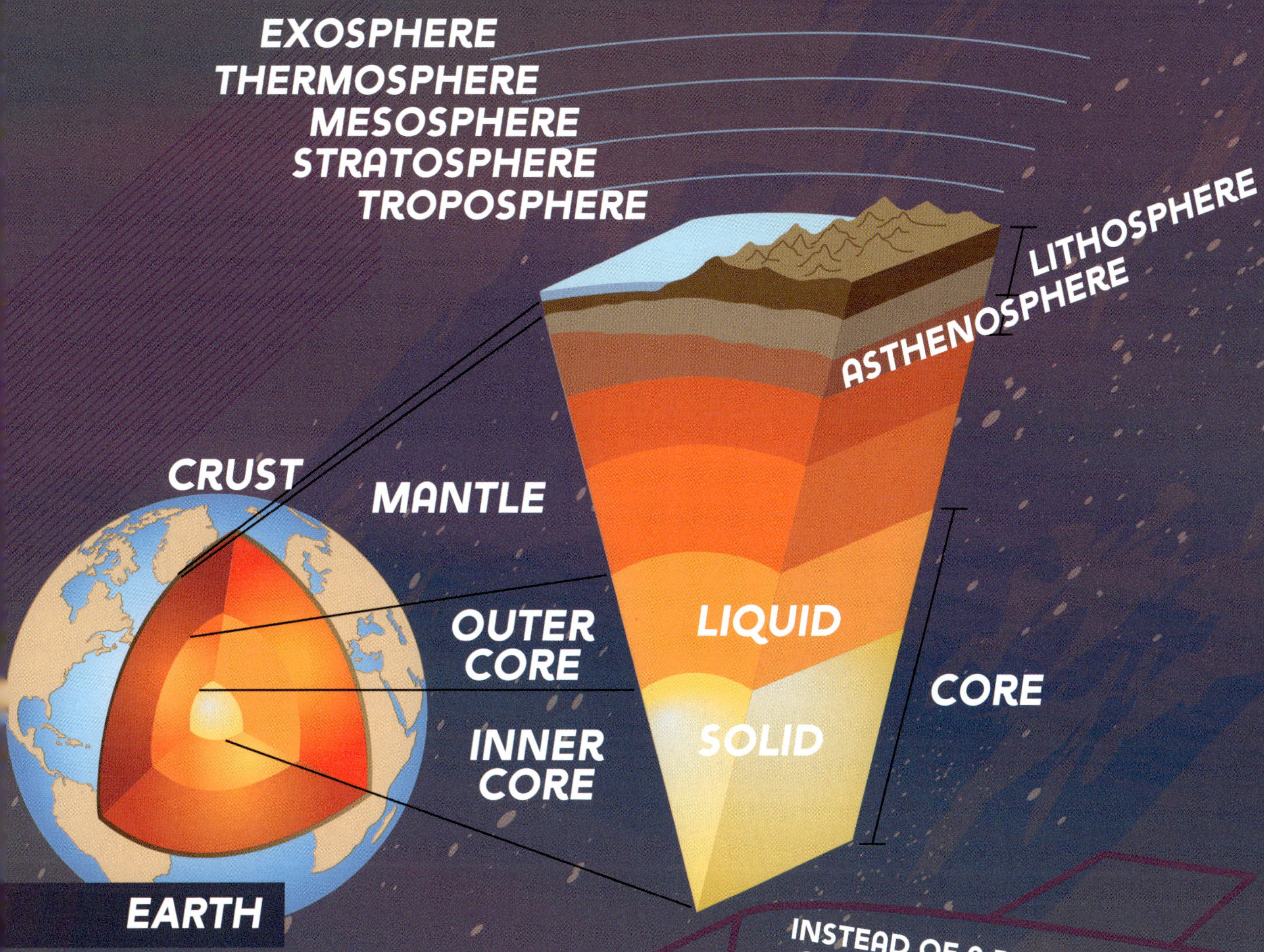

INSTEAD OF A ROCKY CRUST, GAS GIANTS ONLY HAVE AN **ATMOSPHERE** BETWEEN THEIR OUTER GAS CLOUD AND INNER CORE.

FINDING THE RIGHT SPOT

All life on Earth needs liquid water, so this is another thing scientists look for in exoplanets that could be other Earths. This means they need to pay attention to planets in the habitable zone. "Habitable" means able to support life.

If a planet is too close to its star, it'll be too hot, and all the water will boil away. If it's too far away, it'll be too cold, and all the water will freeze. The habitable zone is in between and just right! That's why it's sometimes called the "Goldilocks zone."

OUR SUN IS ACTUALLY VERY SMALL COMPARED TO OTHER STARS, SO ITS HABITABLE ZONE IS ALSO SMALLER.

EXPLORE MORE!

OUR SUN IS A KIND OF STAR CALLED A YELLOW **DWARF** STAR. IT'S SMALLER, COOLER, AND LIGHTER THAN BLUE STARS, BUT BIGGER, HEAVIER, AND HOTTER THAN RED DWARF STARS. THE HABITABLE ZONE IS FARTHER AWAY FROM BLUE STARS AND CLOSER TO RED STARS.

HARD TO FIND

For a long time, scientists thought exoplanets probably existed, but they couldn't prove it. Space is too big, stars are too bright, and planets are too dim. Telescopes weren't powerful enough to see them.

Today, scientists have better telescopes that can find exoplanets through direct imaging. This is when a telescope looks directly at a planet. However, because stars outshine planets and exoplanets are so far from Earth, it only works for planets that are bigger than Jupiter and far away from a star.

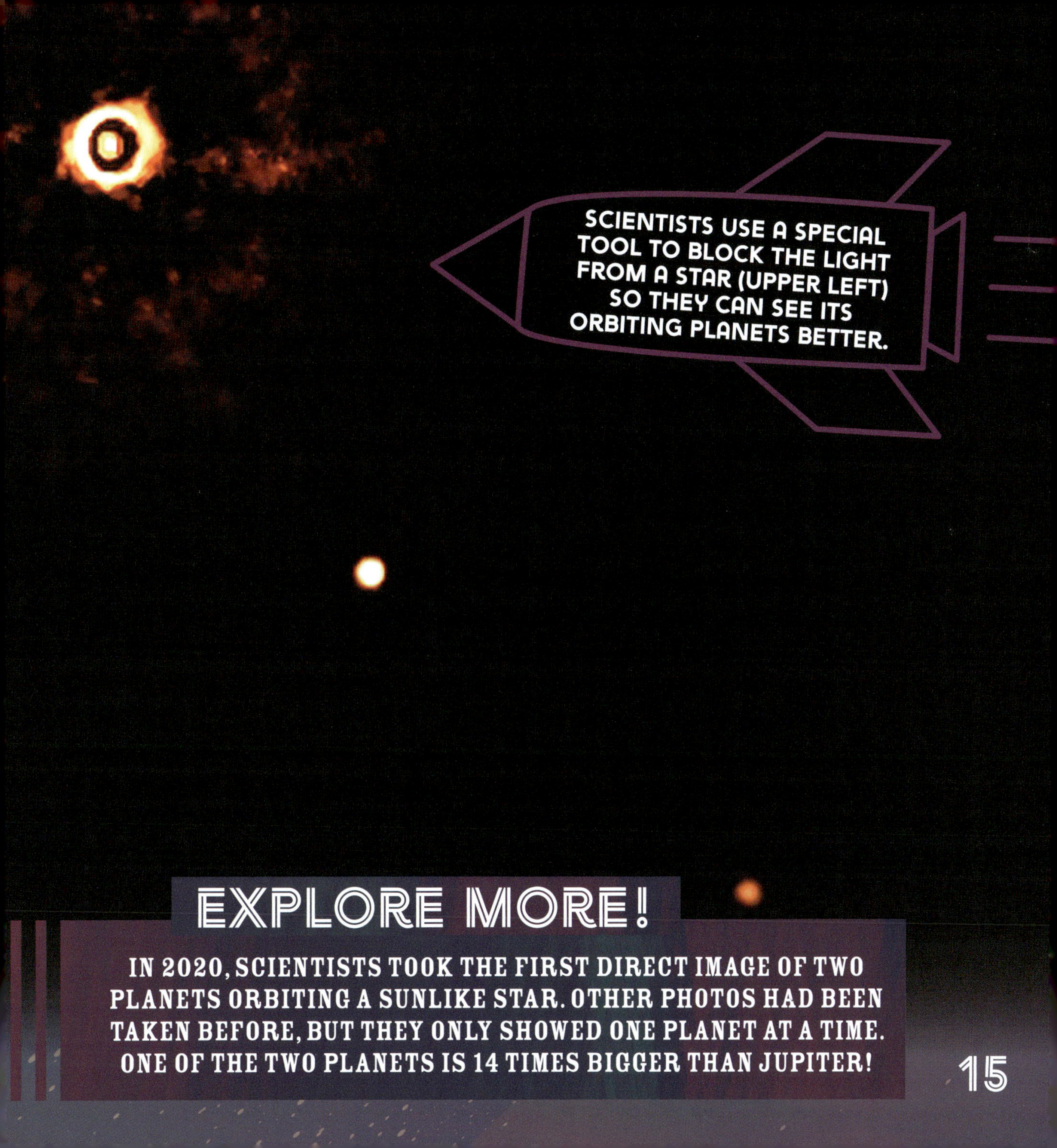

SCIENTISTS USE A SPECIAL TOOL TO BLOCK THE LIGHT FROM A STAR (UPPER LEFT) SO THEY CAN SEE ITS ORBITING PLANETS BETTER.

EXPLORE MORE!

IN 2020, SCIENTISTS TOOK THE FIRST DIRECT IMAGE OF TWO PLANETS ORBITING A SUNLIKE STAR. OTHER PHOTOS HAD BEEN TAKEN BEFORE, BUT THEY ONLY SHOWED ONE PLANET AT A TIME. ONE OF THE TWO PLANETS IS 14 TIMES BIGGER THAN JUPITER!

WHAT CAN DIRECT IMAGING TELL US?

When scientists take a photo of an exoplanet through direct imaging, it isn't clear at all. In fact, the planet just looks like a dot of light. However, this can tell scientists a lot more than we think!

Spectroscopy is a science that studies light. Everything gives off a color. Using special tools, scientists can see these colors. If they see the color of **oxygen** in the light a planet gives off, they'll know there's probably life there. The only reason there's oxygen on Earth is because living things such as plants create it.

THE COMBINATION OF COLORS AN ELEMENT GIVES OFF IS CALLED ITS LIGHT SIGNATURE.

OXYGEN

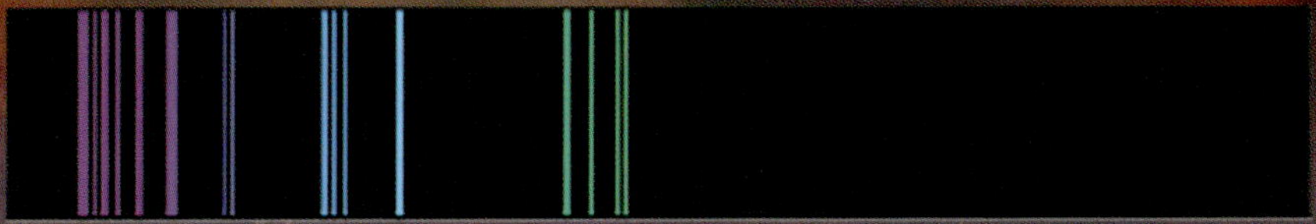

EXPLORE MORE!

THE EXOPLANETS THAT HAVE BEEN FOUND THROUGH DIRECT IMAGING ARE OUTSIDE THE HABITABLE ZONE. HOWEVER, EVEN IF THEY DON'T HAVE LIFE ON THEIR SURFACE, FINDING OUT WHAT OTHER PLANETS ARE LIKE CAN TELL US MORE ABOUT THE UNIVERSE.

LOOKING INDIRECTLY

Most exoplanets aren't found through direct imaging. To find the ones that are too small or close to a star to be seen this way, scientists need to use indirect methods. This means using clues to tell that a planet we can't see exists.

One of the most common indirect methods is watching for a planet's transit—when it passes in front of a star. The planet is too far away for scientists to actually see, but they can measure the amount of light that's blocked. With special tools, scientists can even **detect** a small planet.

EXPLORE MORE!

A STAR'S GRAVITY PULLS ITS PLANETS WHILE THE PLANETS TRY TO BREAK FREE. THIS IS WHAT CAUSES A PLANET TO ORBIT A STAR. IT ALSO CAUSES THE STAR TO "WOBBLE" JUST THE TINIEST BIT. LOOKING FOR THIS WOBBLE IS ANOTHER INDIRECT METHOD.

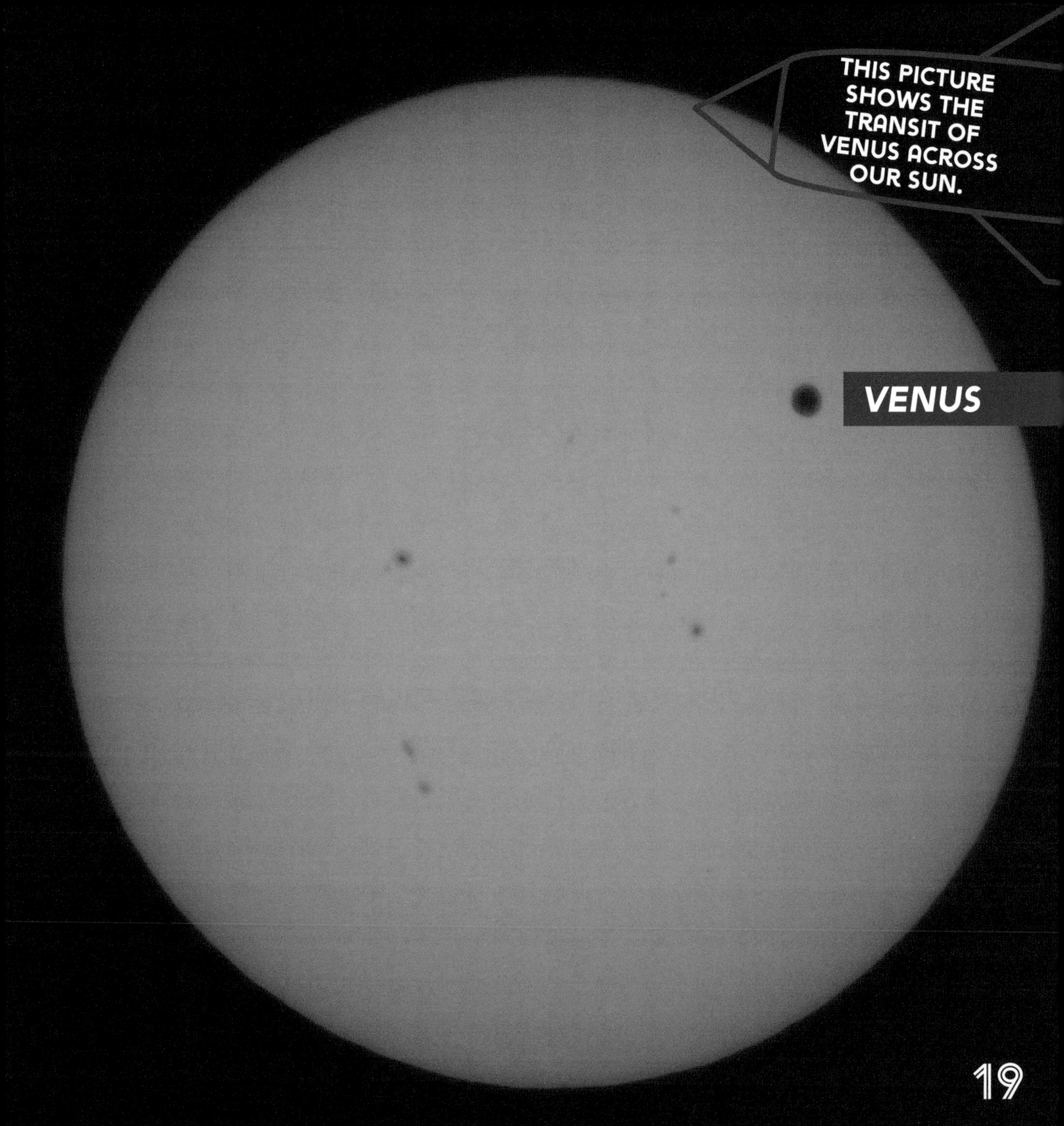

THIS PICTURE SHOWS THE TRANSIT OF VENUS ACROSS OUR SUN.

A HAPPY ACCIDENT

The first exoplanets were found by accident! In 1992, two scientists were studying a pulsar 2,300 light-years away. A pulsar is a star that gives off waves (such as radio waves) in regular pulses, or beats. The scientists noticed that at the same time every few months, the pulsar's pulses were different. The scientists finally figured out that the pulsar had two planets orbiting it. The different pulses happened because of the planets' orbits.

For many years, scientists thought planets that orbit a pulsar couldn't have life on them. Today, some think it might be possible, but we haven't found proof yet.

THIS PICTURE SHOWS THE VELA PULSAR, WHICH IS ABOUT 1,000 LIGHT-YEARS FROM EARTH.

EXPLORE MORE!

NASA IS FINDING NEW EXOPLANETS AT A VERY QUICK RATE. ON MARCH 25, 2022, NASA **CONFIRMED** ITS 5,000TH EXOPLANET. BY APRIL 29, 14 MORE HAD BEEN CONFIRMED. THERE ARE ALSO MANY POSSIBLE EXOPLANETS THAT HAVEN'T BEEN CONFIRMED YET BUT MIGHT BE SOON.

OTHER JUPITERS

Scientists haven't found other Earths yet, but they have found planets they call "hot Jupiters." These are gas giants about the size of Jupiter that orbit their stars in 10 Earth days or fewer. Life as we know it couldn't exist on planets like these.

In the 1990s, most of the exoplanets that scientists found were hot Jupiters. This was because they were the easiest to find with indirect methods. However, as scientists kept looking, they realized there are a lot of different kinds of exoplanets, and hot Jupiters aren't the most common.

EXPLORE MORE!

IN 1995, A SCIENTIST FOUND THE FIRST EXOPLANET THAT ORBITS A SUNLIKE STAR. THIS PLANET IS A HOT JUPITER CALLED 51 PEGASI B. ONE YEAR ON THIS PLANET IS EQUAL TO ABOUT FOUR EARTH DAYS!

THIS PICTURE SHOWS ONE IDEA OF WHAT A HOT JUPITER MIGHT LOOK LIKE UP CLOSE.

SPACE TELESCOPES

Even the most powerful telescopes are limited on the ground. Earth's atmosphere distorts, or changes, light that passes through it. Putting telescopes into space gets rid of that distortion and puts the telescope closer to the things it's observing.

In 1968, NASA **launched** an observatory into space. This was a spacecraft that orbited Earth and had many scientific tools, including telescopes, onboard to allow it to observe objects in space. Scientists who worked together on it later made a group called the Space Telescope Science Institute.

THE SUN LOOKS BLURRY AROUND THE EDGES IN THIS PHOTO BECAUSE IT'S BEING SEEN THROUGH EARTH'S ATMOSPHERE.

CHANDRA

EXPLORE MORE!

SINCE 1970, MANY SPACE TELESCOPES HAVE BEEN LAUNCHED. SOME, SUCH AS ONE NAMED CHANDRA, MEASURE THE X-RAYS AN OBJECT GIVES OFF. THESE X-RAYS ARE ABSORBED, OR SOAKED UP, BY EARTH'S ATMOSPHERE, SO THEY CAN ONLY BE MEASURED BY A SPACE TELESCOPE.

In March 2009, NASA launched the Kepler space telescope. By watching for transits, Kepler found 2,662 exoplanets. One of them was the first planet to be found in the habitable zone of a sunlike star.

In 2018, when Kepler ran out of fuel, NASA shut it down. In 2021, it launched the James Webb Space Telescope (JWST). This telescope measures the heat an object gives off. A very hot rocky planet could be a sign of life.

THE JWST IS THE MOST POWERFUL SPACE TELESCOPE TO BE MADE SO FAR.

EXPLORE MORE!

IN 1990, THE HUBBLE TELESCOPE—THE FIRST MAJOR SPACE TELESCOPE—WAS LAUNCHED. SCIENTISTS USED WHAT THEY LEARNED FROM HUBBLE TO BUILD THE JWST AND MAKE IT MORE POWERFUL. IT WILL BE ABLE TO DETECT OBJECTS UP TO 100 TIMES DIMMER THAN HUBBLE CAN.

STILL LOOKING

In 2014, NASA found the first Earth-sized planet in the habitable zone of its red dwarf star. Scientists think it's probably rocky. They called it Kepler-186f. Other planets also hold promise, and more are being found all the time.

Without getting close to these planets, we can only guess whether they're enough like our own for us—or something like us—to live on. However, it wasn't so long ago that we didn't even know if any exoplanets existed. Science is always moving forward, so we might have the answers soon.

THIS PHOTO SHOWS THE DIFFERENCE IN SIZE BETWEEN OUR SUN AND TRAPPIST-1'S SUN.

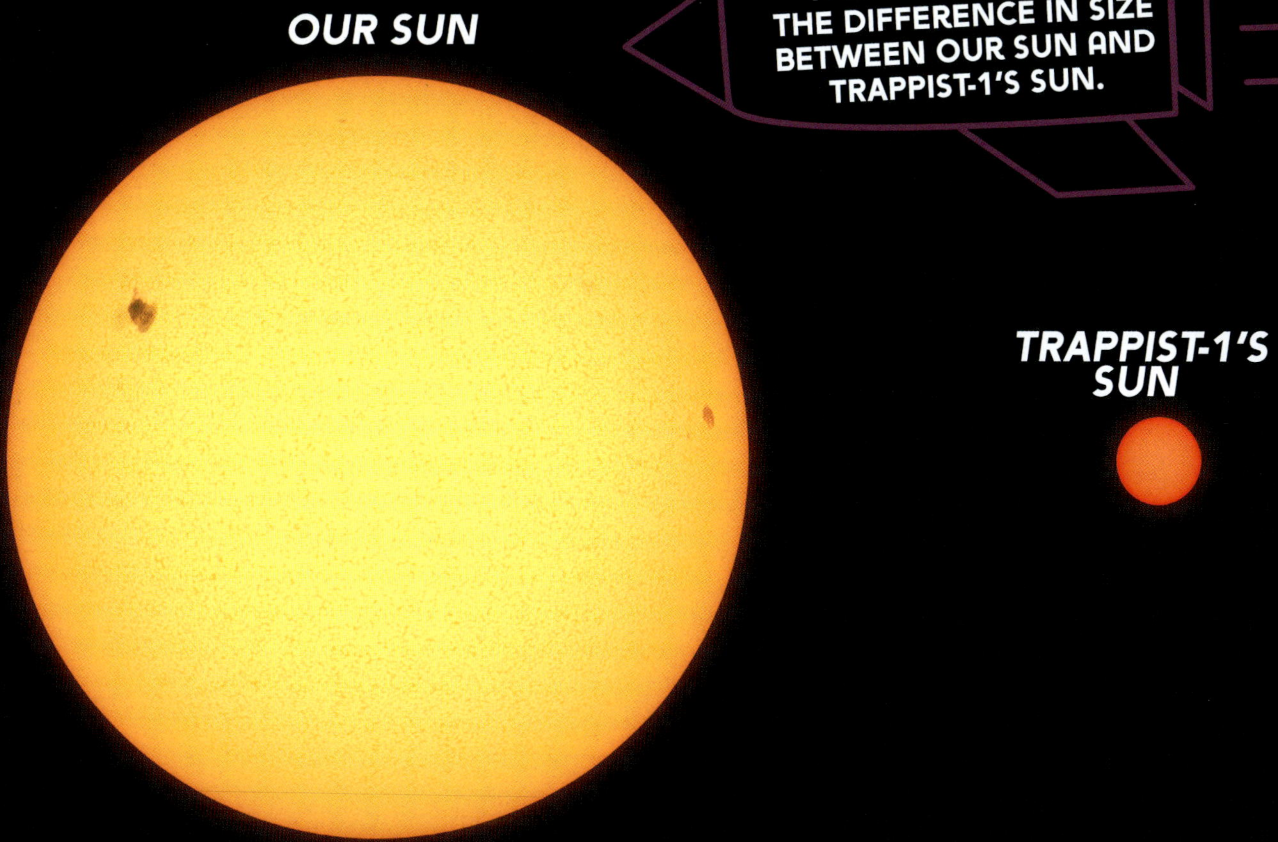

EXPLORE MORE!

THE JWST'S FIRST STOP IS TRAPPIST-1, A SOLAR SYSTEM 40 LIGHT-YEARS AWAY FROM EARTH. THE STAR IN THIS SYSTEM IS A RED DWARF, AND THREE OF THE ROCKY PLANETS THAT ORBIT IT ARE IN THE HABITABLE ZONE.

GLOSSARY

atmosphere: The mixture of gases that surround a planet.

confirm: To find to be true.

detect: To notice or discover the existence of something.

dwarf: Of less than the usual size.

estimate: To make a careful guess about an answer based on the known facts.

evolve: To grow and change over time.

gravity: The force that pulls objects toward the center of a planet or star.

launch: To send out with great force.

observable: Able to be seen.

orbit: To travel in a circle or oval around something, or the path used to make that trip.

oxygen: A colorless, odorless gas that many animals, including people, need to breathe.

universe: Everything that exists.

FOR MORE INFORMATION

BOOKS

Anderson, Amy, and Brian Anderson. *Space Dictionary for Kids: The Everything Guide for Kids Who Love Space*. London, UK: Routledge, 2021.

Lawrence, Ellen. *Looking for Another Earth*. New York, NY: Bearport Publishing, 2019.

Terp, Gail. *The Search for New Planets*. Mankato, MN: Black Rabbit Books, 2019.

WEBSITES

BrainPOP: Exoplanets

www.brainpop.com/science/space/exoplanets

Watch videos and play games to learn more about exoplanets.

NASA SpacePlace

spaceplace.nasa.gov/all-about-exoplanets/en/

Read more about how scientists find exoplanets.

The Nine Planets: Exoplanets

nineplanets.org/kids/exoplanets-planets

Learn more amazing facts about exoplanets.

Publisher's note to educators and parents: Our editors have carefully reviewed these websites to ensure that they are suitable for students. Many websites change frequently, however, and we cannot guarantee that a site's future contents will continue to meet our high standards of quality and educational value. Be advised that students should be closely supervised whenever they access the internet.

INDEX

atmosphere, 11, 24, 25

black hole, 8

direct imaging, 14, 15, 16, 17, 18

gas giants, 10, 22

geocentric model, 5

gravity, 8, 18, 28

habitable (Goldilocks) zone, 12, 13, 17, 26, 29

"hot Jupiters," 22, 23

indirect methods, 18, 22J

Jupiter, 10, 14, 15, 22

light signature, 17

light-year, 7, 21, 29

Milky Way, 8

Neptune, 10

oxygen, 16, 17

pulsar, 20, 21

Saturn, 10

spectroscopy, 16

telescopes, 4, 14, 24, 25, 26, 27

transits, 18, 19, 26

Uranus, 10V

Venus, 19

water, 8, 12